SAE EDGE™ RESEARCH REPORT

Unsettled Domains Concerning Autonomous System Validation and Verification Processes

Fabio Alonso da Silva
ElletroCrafts Aerospace Ind.

EDGE DEVELOPMENT TEAM

Alexandre Gesualdi, *ACG Consulting BR*
Paulo Allemand Donato, *retired from Embraer*
Robert Voros, *Textron Aviation*

SAE INTERNATIONAL®
Warrendale, Pennsylvania, USA

INTERNATIONAL®

About the Publisher

SAE International® is a global association of more than 128,000 engineers and related technical experts in the aerospace, automotive, and commercial-vehicle industries. Our core competencies are lifelong learning and voluntary consensus standards development. Visit sae.org

SAE EDGE™ Research Report Disclaimer

SAE EDGE™ Research Reports focus on topics that are dynamic, in which knowledge is incomplete, and have yet to be standardized. They represent the collective wisdom of a group of experts and serve as a practical guide to the reader in understanding unsettled subject matter. They are not meant to provide a recommended practice or protocol. The experts engaged have contributed their own thoughts and points of view, and these are not the positions of the institutions or businesses with which they are affiliated. A professional writer has collectivized their input; there is no one contributor's perspective being advanced, but rather that of a community of practitioners. SAE EDGE™ Research Reports are the property of SAE International and SAE alone is responsible for their content.

About This Publication

SAE EDGE™ Research Reports provide state-of-the-art and state-of-the -industry examinations of the most significant topics in mobility engineering. SAE EDGE™ contributors are experts from research, academia, and industry who have come together to explore and define the most critical advancements, challenges, and future direction in areas such as vehicle automation, unmanned aircraft, Internet of Things and connectivity, cybersecurity, advanced propulsion, and advanced manufacturing.

Related Resources

SAE MOBILUS® Automated & Connected Knowledge Hub
http://sae.org/Mobilus/Automated

SAE Team

Frank Menchaca, Chief Product Officer
Michael Thompson, Director, Standards, Information and Research Publications
Monica Nogueira, Acquisitions Director
Jill Leonard, Product Manager
William Kucinski, Managing Technical Editor

Printed in USA

ISSN 2640-3536
e-ISSN 2640-3544
ISBN 978-1-4686-0127-5

To purchase bulk quantities, please contact: SAE Customer Service

E-mail: CustomerService@sae.org
Phone: 877-606-7323 (*inside USA and Canada*)
724-776-4970 (*outside USA*)
Fax: 724-776-0790

Visit the SAE International Bookstore at books.sae.org

About the Editor

Fabio Alonso da Silva is currently the Head of Systems at ElletroCrafts Aerospace Ind., a mobility startup tasked with developing an urban air mobility vehicle or "flying car." Additionally, da Silva is pursuing an MBA at the Warwick Business School/University of Warwick, UK.

With more than 20 years of experience in systems engineering, development, integration, and product sustainment, da Silva has worked at Embraer, Bombardier, Gulfstream, and recently Zoox, an automated driving system company. His professional journey has focused on development of fly-by-wire aircraft systems and autonomous systems for autonomous vehicles [1].

Da Silva is dedicated to the deployment of safe solutions for autonomous systems and made his mission to contribute to the discussion of the creation of a framework that would support obtaining safety clearance of such systems. He combines perspectives from technology, business, and vehicle certification.

contents

Atosan/Shutterstock.com

Unsettled Domains Concerning Autonomous System Validation and Verification Processes

Abstract

The Federal Aviation Administration (FAA) and the Department of Transportation's (DOT's) National Highway Traffic Safety Administration (NHTSA) face similar challenges regarding the regulation of autonomous systems powered by artificial intelligence (AI) algorithms that replace the human factor in the decision-making process. Validation and verification (V&V) processes contribute to implementation of correct system requirements and the development life cycle - starting with the definition of regulatory, marketing, operational, performance, and safety requirements. The V&V process is one of the steps of a development life cycle starting with the definition of regulatory, marketing, operational, performance, and safety requirements. They define what a product is, and they flow down into lower level requirements defining control architectures, hardware, and software. The industry is attempting to define regulatory requirements and a framework to gain safety clearance of such products. This report suggests a regulatory text and a safety and V&V approach from an aerospace engineering perspective assessing the replacement of the human driver from the decision-making role by a computational system. It also suggests an approach where aerospace guidelines can be used alongside the International Organization for Standardization (ISO) standard ISO 26262 in order to define a viable and valuable framework for autonomous systems safety clearance (or certification).

NOTE: SAE EDGE™ Research Reports are intended to identify and illuminate key issues in emerging, but still unsettled, technologies of interest to the mobility industry. The goal of SAE EDGE™ Research Reports is to stimulate discussion and work in the hope of promoting and speeding resolution of identified issues. SAE EDGE™ Research Reports are not intended to resolve the issues they identify or close any topic to further scrutiny.

FABIO ALONSO DA SILVA
ElletroCrafts Aerospace Ind.

Edge Development Team

Alexandre Gesualdi, *ACG Consulting BR*
Paulo Allemand Donato, *retired from Embraer*
Robert Voros, *Textron Aviation*

ISSN 2640-3536

Introduction

This SAE EDGE™ Research Report discusses important aspects related to the V&V processes of the AI system and drive-by-wire system (DBWS) currently utilized in self-driven vehicles from the perspective of aerospace methodologies and processes. (The aerospace field is also working to solve the challenge of developing autonomous flying vehicles.) Some references suggest that a possible way to address this challenge is to see how aerospace deals with the development of highly integrated and complex systems [2, 3].

From autopilot systems to full fly-by-wire flight controls, aircraft have embodied autonomy with varying degrees of complexity and integration. However, this aviation implementation does not involve the same type of complex operating scenarios experienced in the automotive field. But these two fields do share a common challenge in the lack of regulation regarding the design of this high level of complex autonomy.

Both industries are looking forward to defining those regulations which will establish the design directions and processes for certification.

Basically, the main challenges to be solved by replacing the human driver (or pilot) with an autonomous system are

- Showing that it will obey the ground or air traffic laws
- It will perform better than a pilot or driver under any condition
- It is safer than a human driver or pilot

Being safer is an undefined concept. For simplicity, we will consider that the autonomous systems for a vehicle must be ten times better than a human driver. Considering those three statements might be used to guide the design, we should understand their implications in the development plan and activities.

Both fields have their own standards, Aerospace Recommended Practice (ARP) 4754A and ISO 26262, to guide the requirement-based design utilizing the V&V process [4]. The ARP is dedicated to product and systems whereas the ISO also covers "items" or software and hardware.

In aviation, items are addressed in the application of requirements from ARP4754A through the software process described in the Radio Technical Commission for Aeronautics (RTCA) DO-178 and airborne electronics hardware process RTCA DO-254 [5, 6].

ARP4754A is more detailed and provides resources for solving this challenge; however, both of these guidelines have similarities between them. The RTCA DO-178, RTCA DO-254, and several FAA articles and papers can also be of great value for software and hardware design.

We cannot blindly assume that adopting the aerospace framework is just enough without considering the automotive approach and others. This challenge is solved by understanding how each can be used, merged, and/or tailored, or how selected parts of each one would best fit in a new framework. This article considers those guidelines in conjunction with ISO 26262. Nevertheless, the aerospace frameworks have been used for such a long time in the development of highly integrated and complex systems, such as fly-by-wire, that their credibility is now a given. This strategy intends to add value and confidence in the development of autonomous systems.

Of particular importance is the development, validation, and verification of software containing machine-learning and decision-making algorithms.

This SAE EDGE™ Research Report accounts for the points of view of different collaborators working in the aerospace and automotive fields. The intention is to suggest an approach that tailors the best and most appropriate processes, tools, and solutions from both fields that give a solid direction for certification of an autonomous system applied to self-driving vehicles.

Note: SAE EDGE™ Research Reports are intended to identify and illuminate key issues in emerging, but still unsettled, technologies of interest to the mobility industry. The goal of SAE EDGE™ Research Reports is to stimulate discussion and work in the hope of promoting and speeding resolution of identified issues. SAE EDGE™ Research Reports are not intended to resolve the issues they identify or close any topic to further scrutiny.

State of the Industry

Different products and business models are sources for different requirements that will go through V&V processes where specific approaches to verification, mainly by testing, might be necessary for each level. Automated driving system economics, average utilization, fatalities, operational costs, ownership, and development of automated driving systems is explored in some detail in the SAE EDGE™ Research Report "Unsettled Technology Areas in Autonomous Vehicle Test and Validation." [7].

Product Types There are different types of vehicles in development, testing, or early operation with different types of configurations in terms of sensors, computers, and automotive systems. Generally, radar, LiDAR, and cameras are used, and their information is merged to create an interpretation of the three-dimensional (3D) world around the vehicle to detect and identify objects or "agents." Sensor types and the number of sensors may vary by vehicle type.

Propulsion systems for automotive platforms can also vary, but hybrid and electric drivetrains dominate the market. Waymo uses an internal combustion engine and electric hybrid van: the Chrysler Pacifica Hybrid (Figure 1) [8]. Tesla produces electric vehicles (Figure 2), Cruise [9] uses a Chevrolet Bolt electric vehicle (Figure 3), Zoox uses an electric vehicle, and Uber uses a Volvo XC90 hybrid sport utility vehicle (Figure 4).

FIGURE 1.

Sundry Photography/Shutterstock.com

FIGURE 2.

franz12/Shutterstock.com

FIGURE 3.

Sundry Photography/Shutterstock.com

FIGURE 4.

Sean Leonard/Shutterstock.com

The company Nuro [10] developed a small vehicle to deliver groceries, and it is already in operation. Kodiak [11] and Starsky [12] (Figure 5) are introducing autonomous systems to the trucking segment.

The complexity of each autonomous vehicle reflects the operational requirements selected according to their business strategies.

Similarly, in the aerospace there are companies, such as Uber [13], Bell Helicopter, and Nexus [14], that are investing in their own versions of embedding autonomous systems that will replace human pilots. (Figure 6) [15, 16]

Market Segments The retail market supplies autonomous solutions embedded in cars, such as Tesla, BMW, and others. The mobility services segment offers a fleet of vehicles for ride-sharing applications, such as Waymo, Cruise, and Zoox. The truck segment (goods transportation in general) is also introducing autonomous systems with different operational profiles. These different products in their market segments all present different levels of autonomy [17] and different levels of design maturity.

Each one of these examples might follow the definition of the high-level requirements captured from different areas, such as marketing, operation, safety, regulatory, maintenance, customers, etc. (requirement-based design suggested in ARP4754A). Together they define the product and services, and the development process (requirement-based design) captures and defines what those high-level requirements are, breaking them down into other layers of more detailed requirements. These will proceed through V&V processes to certify that the correct requirements are in place, and they correspond to what has been defined. In further sections we will visit in detail this whole process and how V&V represents important steps to obtaining "safety clearance" needed for "certification" (or "homologation" as per ISO 26262).

Generic Architecture The level of automation has increased in vehicles, and traditional mechanically actuated automotive systems have been replaced by electromechanical ones with embedded electronics and software. Automotive systems have their own control boxes performing safety

FIGURE 5.

Starsky_Robotics/Shutterstock.com

FIGURE 6.

James Mattil/Shutterstock.com

critical functions; they interface with the driver and other systems in order to perform their intended function (Figure 7).

At a minimum, the AI system and DBWS will use LiDAR, radar, and/or cameras. Those sensors acquire information about the vehicle's surroundings and support perception and prediction functionalities, which categorize surrounding objects as either static or dynamic. In application, this means that the vehicle can differentiate between a pedestrian walking across the street and a stationary item like a building, and estimate where dynamic objects might be or if they will interfere with the vehicle's defined trajectory. The AI computer merges all information and defines trajectories that the drive-by-wire computer (DBWC) reads and transforms into commands for the automotive system controllers to execute actions like braking, accelerating, and steering. These computational systems, sensors, and software should be designed and defined to have the same or better performance than a human driver under normal conditions, and also when faced with unsafe conditions, inclement weather, variable lighting conditions, road traffic, and dynamic surroundings (pedestrians, jay walkers, etc.) under normal and/or degraded operational modes. V&V efforts must establish that this complex system has clear requirements and that it operates as designed. For aerospace systems, these efforts are based on guidelines established in ARP4754A.

Autonomous System Sometimes the boundaries between an automatic and autonomous system get blurred or even overlap. For the sake of this report, an automatic system is responsible for maintaining waypoints and is dedicated to a reduced number of functions (e.g., a braking or steering control system in a vehicle). An autonomous system controls multiple automatic systems, reading their information and issuing commands or new waypoint for each automatic system to adapt to. Autonomous systems replace the human factor in making decisions, situational awareness, and issuing commands.

With that in mind, at any point in the development and technical analysis of a vehicle, all responsibilities and requirements normally filled by a human are transferred to the autonomous system. However, different companies may choose different architectures and functional allocations to support autonomous system development (Figure 8).

Automatic systems are complex in their own right, their performance depends on data from other systems to perform safety-critical functions. The autonomous system

FIGURE 7. **A generic architecture of an autonomous vehicle, in this case electrically driven. This figure captures the different components of a hypothetical autonomous vehicle.**

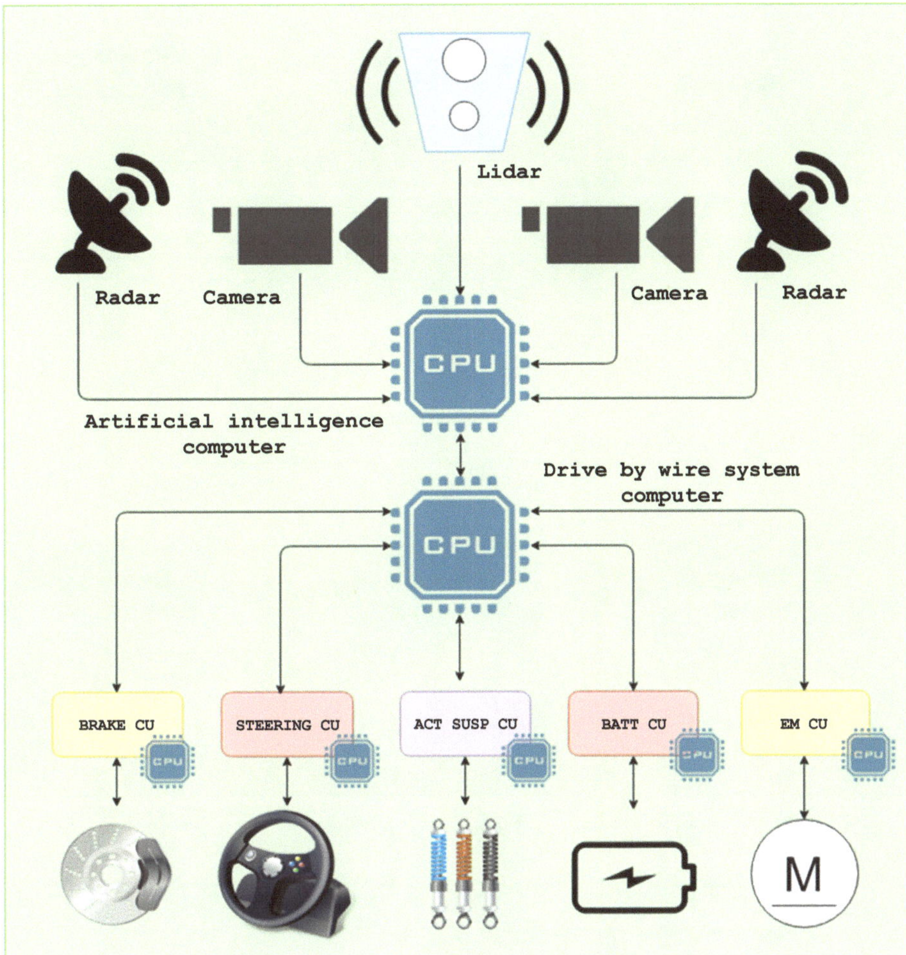

also performs critical safety functions and depends on data from all other automatic systems and sensors, classifying it as a highly integrated and complex system that requires a demanding and extensive analysis. Current aerospace guidelines bring more detailed guidance than the automotive-based ISO 26262, and it can contribute to the overall definition of a development process tailored to fit the autonomous system development and safety clearance.

Highly Integrated and Complex Systems Development Life Cycle To discuss the characteristics and challenges of V&V processes for an autonomous system applied to a ground vehicle, the whole V&V process needs to be viewed in the context of a highly integrated and complex-systems-development life cycle. To achieve this, the characteristics of such a system must be identified, their unique requirements and their type (e.g., safety, operational, etc.) must be determined. There are levels of definitions (e.g., product, system, item, etc.), and the activities for V&V are specific to each level and will be discussed further in this report.

Development Process Life Cycle. The development life cycle of a product and its systems can be split in three phases: concept, development, and production/operation. In aerospace, "product" refers to aircraft; in our report, autonomous vehicle is understood as the overall product and the system under analysis is the autonomous one. (Figure 9)

Requirement-Based Development. ARP4754A suggests a requirement-based development in which requirements are initially defined at the high level or product level, usually classified into distinct types, or sources, as originating from operations, marketing, performance, safety, etc. They all define the overall product functionalities and characteristics which are further refined, or broken down, into product-level

FIGURE 8. **Hypothetical System Control Architecture. Additional sensors provide needed measurements to supplement automated (or automatic, per this figure's terminology) systems to fulfill safety, operational, and/or safety requirements for the collective autonomous system.**

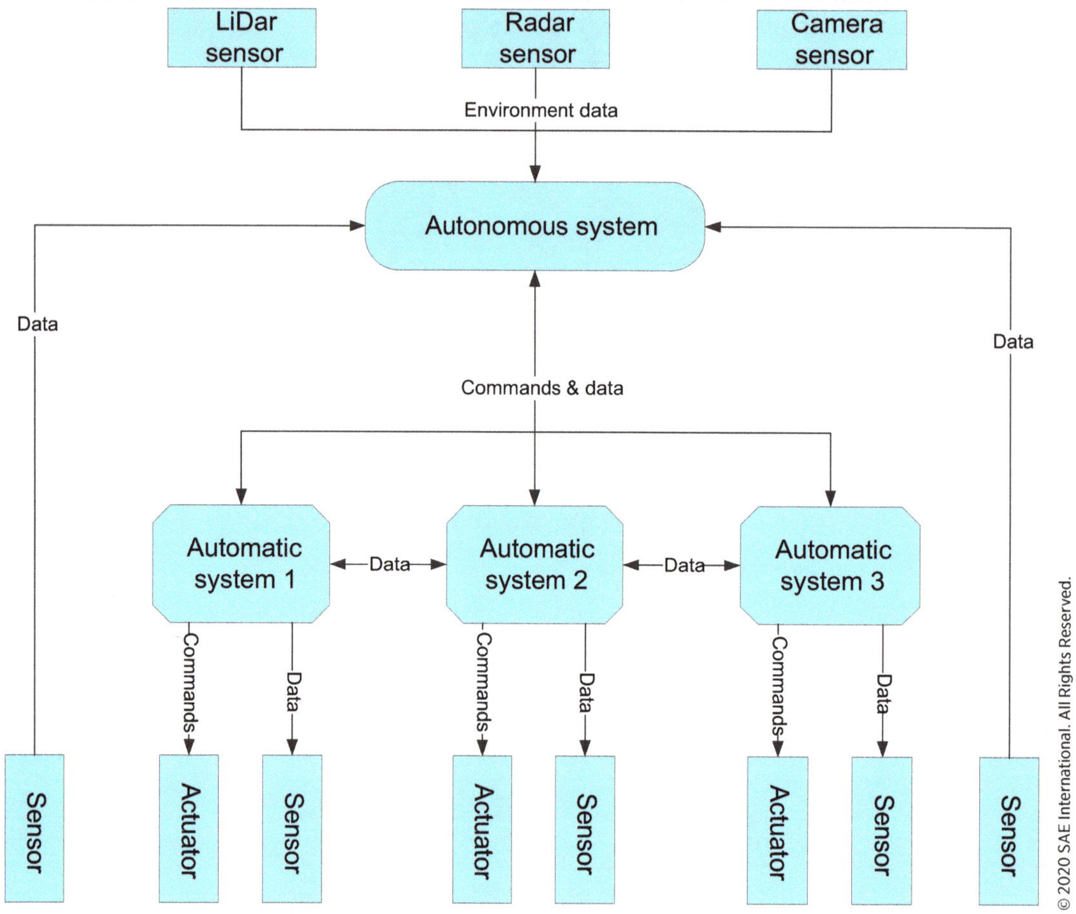

FIGURE 9. **Development Life Cycle (ARP4754). The conceptual phase establishes the product definitions and high-level characteristics. The product is matured during the development phase and is readied for production/operation.**

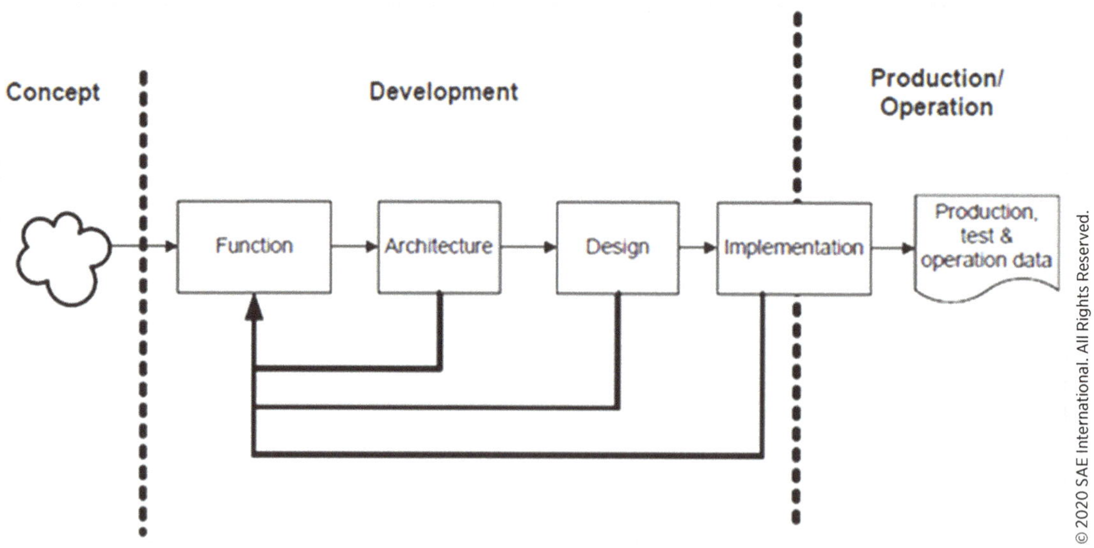

FIGURE 10. **Requirements Flow (based on ARP4754 contents).**

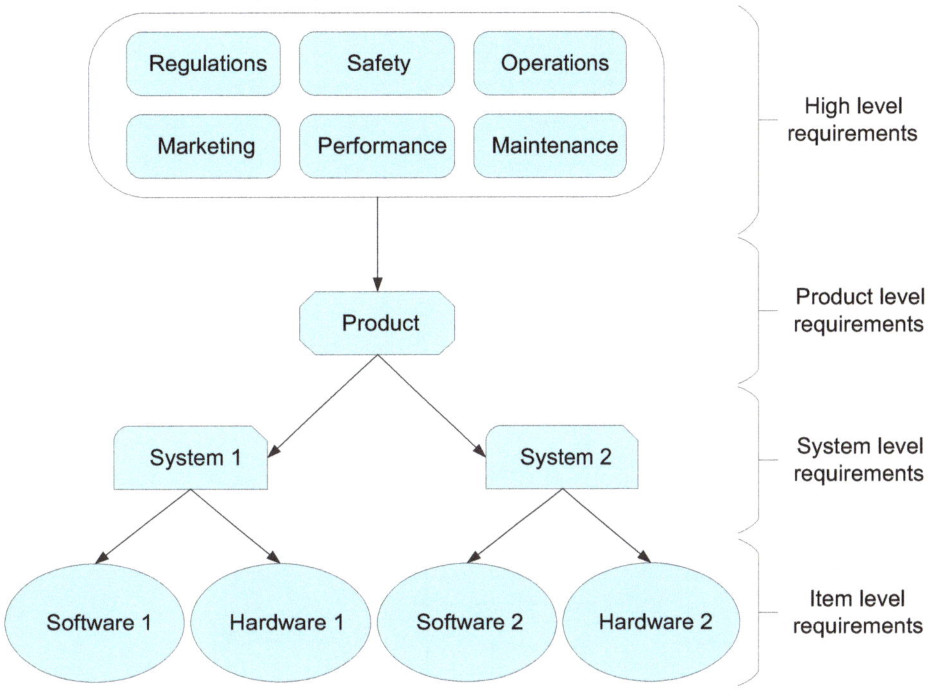

requirements, system-level requirements, and item-level requirements, as seen in Figure 10. This report will focus mainly on regulatory and safety requirements since the unsettled aspects under analysis derive from them.

Functional Development. Figure 11 shows an aircraft function implementation process. The model includes multiple system development processes. Each system-development process can consist of multiple item development processes. There are certain integral processes that take place repetitively during each of the development activities, such as safety assessment, validation, and verification. This framework is suggested for the functional development of the autonomous system.

Unsettled Domains Concerning Autonomous System Validation and Verification Processes

This SAE EDGE™ Research Report considers the following areas to be unsettled domains in autonomous system V&V:

- Regulation requirements to define autonomous system design, what they should be, and how to document compliance
- Safety assessment process, the ARP4761 [18] and ISO 26262 [19] approaches
- Control architecture and hardware
- Software V&V challenges

Design Regulations

The lack of proper design regulations lead manufacturers to pursue various approaches for certification or safety clearance. This can seriously affect the development process and safety. The risk of not having proper and standard guidance results in the neglect of important aspects in the design of autonomy, which has consequences for the development of the technology and its credibility. There are initial efforts in that direction [20], from both automotive and aerospace industries, even though some aspects of technology used in autonomous vehicles remain unsettled [21].

The FAA - via Federal Aviation Regulations (FARs) - publishes regulations for each type of aircraft category and usage. Rulemaking covering the design of autonomous systems is expected; FAA has publicly announced their intention to speed up their processes for certification of aerospace products to cope with new and breakthrough technologies [14, 15]. NHTSA has released an "advance notice of proposed rulemaking" (ANPRM) [22] to gather contributions for the definition of its pilot program to define a viable framework for development and testing of autonomous vehicles. Another

FIGURE 11. **Integral Processes (ARP4754).**

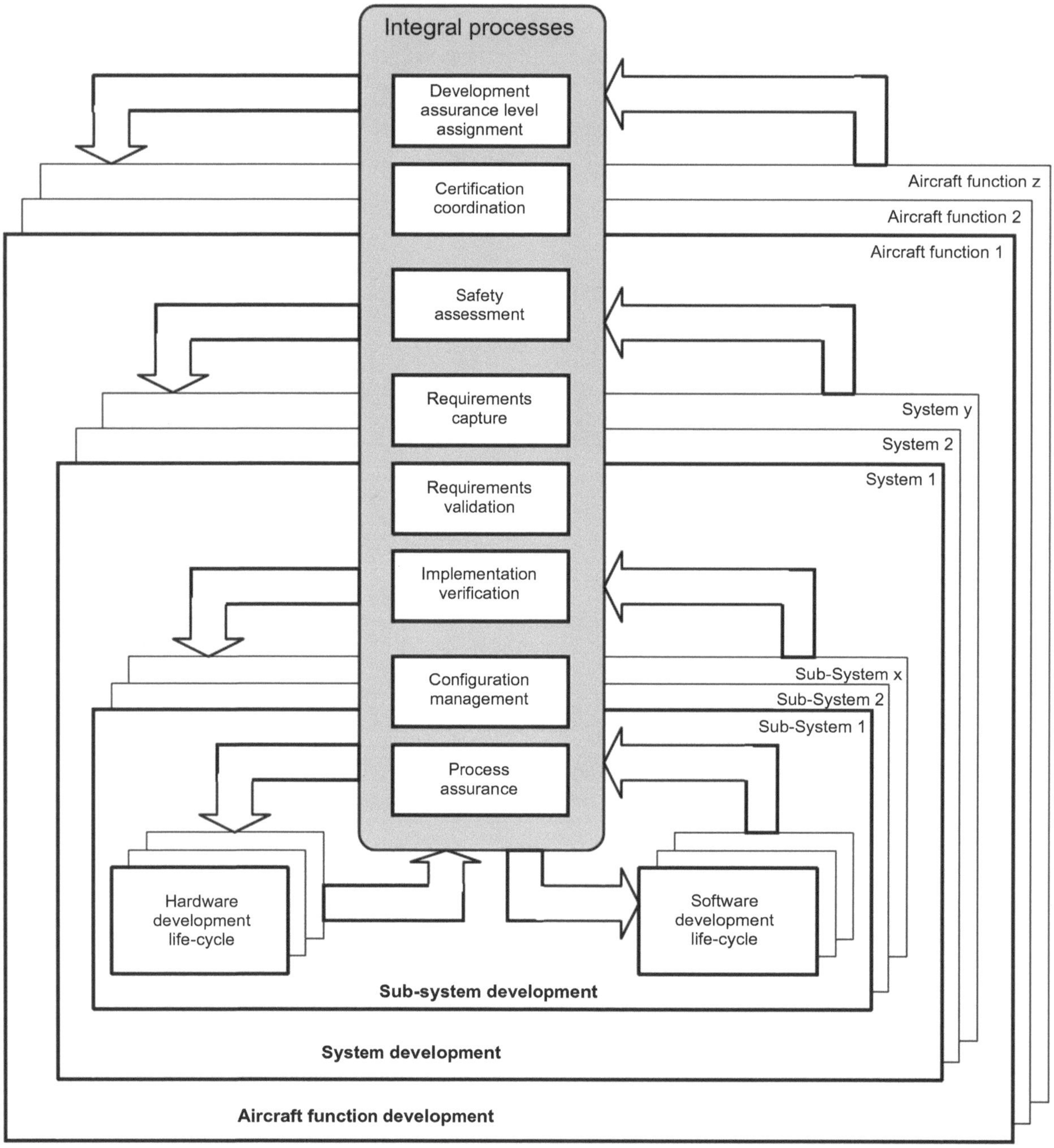

related challenge is the lack of regulations for operation and testing of autonomous vehicles in each US state; however, this is not addressed in this report [23, 24].

Autonomous System Regulatory Requirements

In order to validate and verify an autonomous system requirement, we need to know what it is and where it is coming from for perspective and context. Ideally, ANPRM (when published) will define the processes and framework for the safety clearance and deployment of autonomous vehicles. The V&V process will play an important role in this.

Figure 12 depicts the flow from high-level sources of requirements to define the lowest level of requirements at the item level. As any other design requirement, the autonomous system is ultimately affected by the definition of the high-level requirements. The high-level

FIGURE 12. **High-level sources of requirements define the lowest level of requirements at the item level.**

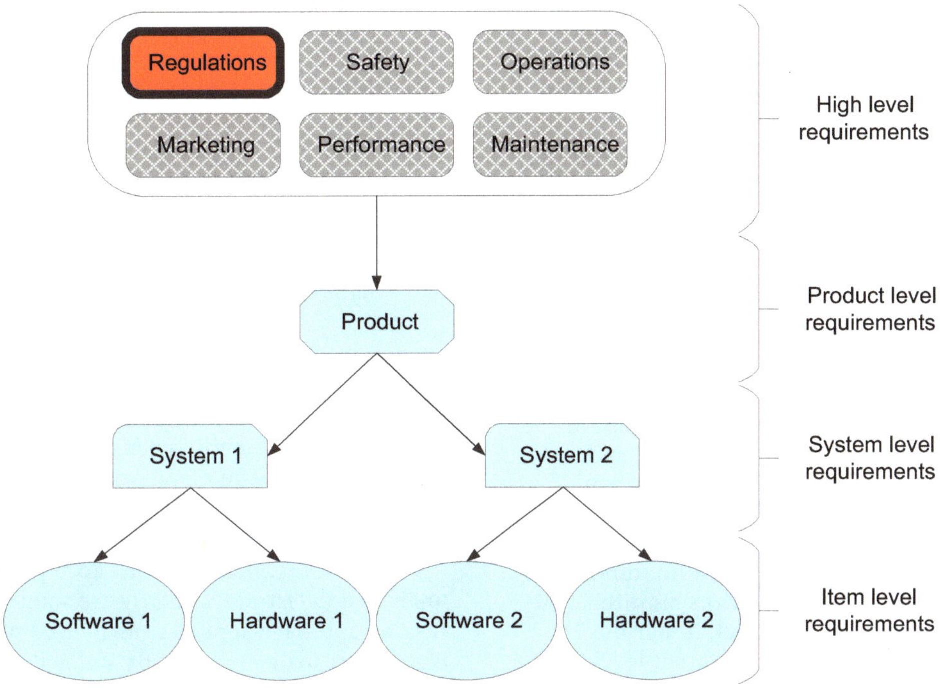

requirements are defined as pertaining to six macro areas of regulations, safety, operations, marketing, performance, and maintenance.

Suggested Regulatory Text Regulations are intended to guide the design of products, systems, and items. The following paragraphs suggest three points that might become the supporting pillars for autonomous system design.

1. Any system, or systems, in an autonomous vehicle/self-driven vehicle replacing the human driver in its functions of driving a vehicle and making decisions has to be shown to comply with all ground traffic rules related to any given operational design domain (ODD) under any given scenario, by proper means, which can be defined by, but not limited to, analysis and/or testing, or any other means, according to the scenario's severity.
2. Any system, or systems, in an autonomous vehicle/self-driven vehicle replacing the human driver in its functions of driving a vehicle and making decisions **has to be shown to have an equal or better performance than a human driver maintaining the safety margins in any given** ODD under any given scenario, by proper means, which can be defined by, but not limited to, analysis and/or testing, or any other means, according to the scenario's severity.
3. Any system, or systems, in an autonomous vehicle/self-driven vehicle replacing the human driver in its functions of driving a vehicle and making decisions has to be shown to be "several times" safer than a human driver in which he or she has caused an incident or fatal accident in any given ODD under any given scenario, by proper means, which can be defined by, but not limited to, analysis and/or testing according, or any other means, to the scenario's severity.

Since this report was developed with an aerospace engineering perspective, this passage reads like FAR. If the terms "human driver" and "ODD" are respectively replaced by "pilot" and "flight envelope," those three bullets can also be considered as possible aerospace regulations.

Those paragraphs have some core elements of text with some variations in them and they can be analyzed in parts:

- **"Any system, or systems, in an autonomous vehicle/self-driven vehicle replacing the human driver in its functions of driving a vehicle and making decisions [...]"** This statement addresses the responsibilities of the human driver mentioned above such as awareness, decision, and action. Those three main responsibilities must be put in context of each of the core demands in each paragraph. For example, the first paragraph could pertain to following traffic rules. Awareness decision, and action imply monitoring, control, and decision-making algorithms and actuation when translated to lower-level item requirements.

- **"... under any given scenario, by proper means, which can be defined by, but not limited to, analysis and/or testing, or any other means, according to the scenario's severity."** This part of the statement defines two important things: one is the relation to scenarios and the second is what option(s) to use in showing compliance. This can include methods like analysis, testing, and others depending on the scenario's criticality (failure condition criticality in the aerospace context) - all of which are properly defined in the verification plan. As it will be seen later, scenarios also form the basis for requirements definition. It is also important to maintain safety levels even with degraded autonomous or automotive systems. This requires monitoring the modes of operation (normal, under no failures, and degraded, with different levels of failures).
- **"[...] has to be shown to comply with all ground traffic rules related to any given ODD [...]"** The autonomous systems has to obey the traffic laws; however, this cannot be looked at simply as obeying the right of way or stopping at the red light. Together with the scenarios they define (safety) behaviors. Scenario relates to all elements or objects (either static or dynamic) found in the environment of concern. The normal operation and fulfillment of those objectives should happen in any ODD. ODD restrictions due to system degradation are also important. The autonomous system must comply with the applicable requirements even in those degraded-conditions cases.
- **"[...] has to be shown to have an equal or better performance than a human driver maintaining the safety margins in any given ODD [...]"** This statement directly addresses the awareness, decision, and action features or how those interact together to deliver an equal or better performance than a human driver. This will break down in time delays and processing capabilities of sensors, processors, and actuators. Much of the credit taken from a human presence is made in the Functional Hazard Analysis (FHA) in aerospace and Hazard Analysis and Risk Assessment (HARA) in the automotive world. Identifying the points where credit is taken from humans indicates what requirements the autonomous system must show to be in compliance with.
- **"[...] has to be shown to be "several times" safer than a human driver in which he or she has caused an incident or fatal accident in any given ODD [...]"** The whole idea behind autonomous vehicles is that they can be safer than a human driver and significantly reduce the number of traffic deaths. However, no metric defines what that limit should be. What number makes sense? How would it translate to product, system, and item designs? Is it achievable? We believe that autonomous systems will reduce the incidents in traffic by some order of magnitude, also we understand the impact on design and how this can be achieved.

Means of Compliance In the aerospace sector, means of compliance (MoC) describes how each product, system, or item feature complies with the regulatory requirement it is trying to satisfy. In many cases, the MoC is performed via testing or the requirements are verified through testing. For example, the bulk of 14 CFR 25.1309 [25] shows compliance via safety assessment and the V&V process, supported by analysis and testing. Similarly, we want to say that the verification approach applied to an autonomous system to show compliance to the "regulations" defined above, at a minimum, must be tested as well.

The point is, then, how to satisfy the requirements derived from the three (3) "regulations" and how to verify them (or find compliance).

To comply with all traffic rules, the autonomous vehicle has to be exposed to all situations in which the correct performance of every traffic rule is confirmed. In order to do that, the product has to have an idea of its surroundings (traffic signs, traffic lights, pedestrians, etc.), which implies reading the environment. It also needs to move at defined speeds. It has to do all of that correctly, with integrity, keeping within safety margins. At the autonomous system level, the need for a sensing system, processing performance, and interface to other automotive systems is implied. Requirements for integrity of commands and data are flowed down to hardware architecture and its reliability and software processes. The software will develop the required functionalities to meet required behavior.

In Figure 13, we see three examples of possible scenarios in which the autonomous vehicle (blue car) has to deal with other cars, pedestrians, and road signs.

Meeting the or exceeding the performance of a human driver requires knowing the average numbers assumed for humans, from becoming aware of a situation, processing, and reacting. This also needs to account for signal speed from sensors and processors (down to actuators for steering, brake, etc.). In the autonomous system detection, this could be defined as processing the information and commanding the brakes and steering system within the time limits without losing control of the vehicle. To comply with the requirement for being several times safer than a human driver, the autonomous vehicle must be submitted to specific selected scenarios in which a human driver has failed to maintain control of a vehicle and ultimately caused an accident.

FIGURE 13. **Scenarios Examples**

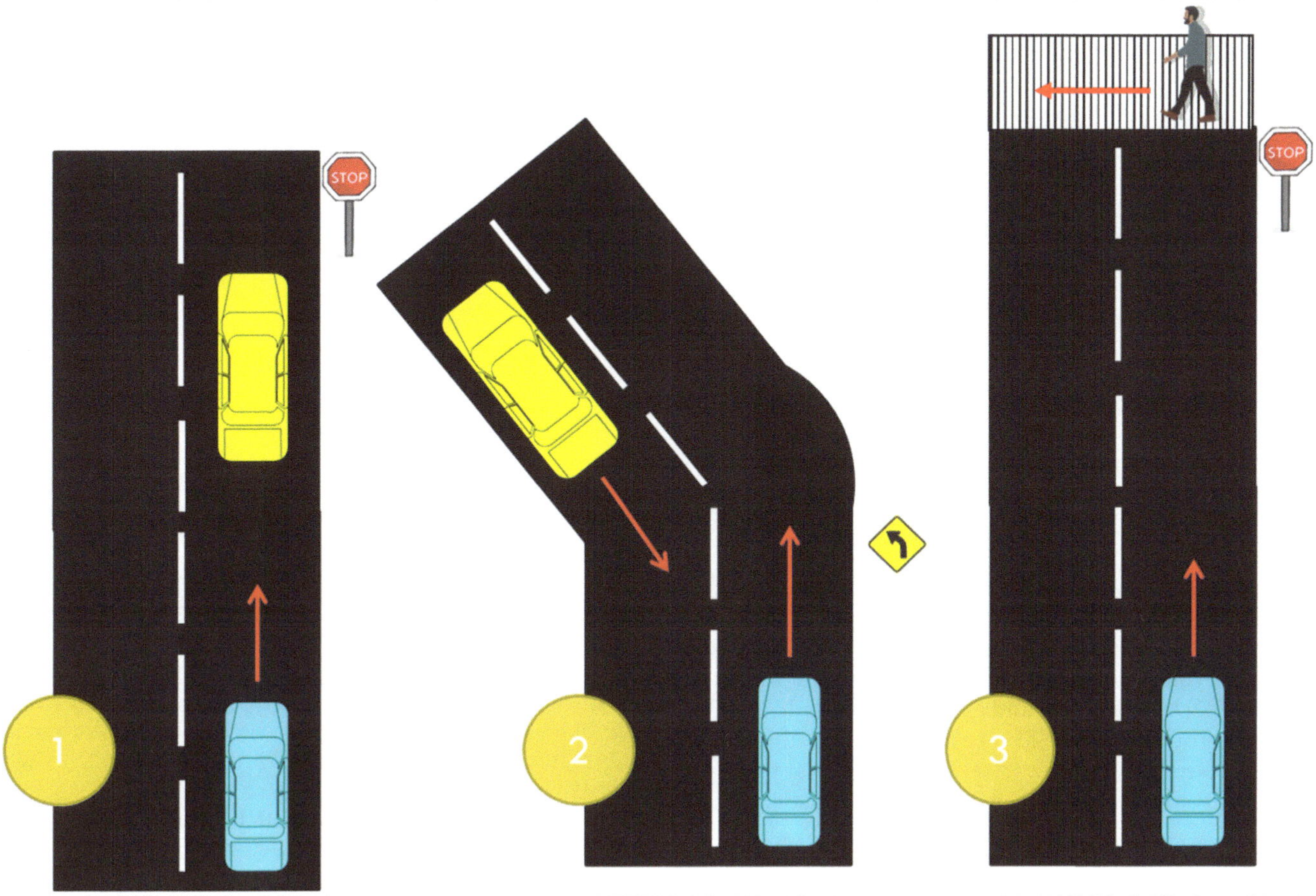

The main methodology to make sure the product, system, and items comply with the requirements derived from the three regulations and scenarios is verification by testing. This can be achieved in three ways [26, 27].

- Simulation (lab testing): This implies in either injecting simulated data directly into the autonomous system computers (AI computer and DBWC) to check each scenario and their enormous variations or performing data injection at the level of sensors to measure the end-to-end integration (from sensors to command outputs or end-effectors). In aerospace there are the "iron birds," which similarly test the end-to-end integration (and partial performance) and functionalities of all systems. Integration labs exist to test electronic equipment (without mechanical elements). This concept can also be used in the autonomous vehicle development. Test labs are useful for risk mitigation and, depending on its parity level to the final product, can be used in formal safety clearance testing before the eyes of certification authorities. Here normal operation and all levels of degradation can be verified.
- Test-Track Testing: A test track is a closed control environment where the performance requirements can be verified (e.g., braking times, steering rates, active suspension stability), and critical scenarios can also be verified with the use of test dummies. This requires the use of a prototype vehicle that is close to the final product configuration (which can also be used to support a certification campaign and capability testing). There are different companies supplying equipment for this type of testing. This would relate to a flight test campaign in the aerospace world, where either ground or flight tests are performed with prototypes with the objective to test functionality and performance. For autonomous vehicles, normal operation and selected levels of degradation can be verified. It is up to each company to define what levels of degradation to test as some can result in severely decreased safety margins.

- Road Testing (city testing): Testing on freeways and urban streets is the ultimate step in the verification process of the autonomous vehicle. In the verification plan and test procedures, describing goals (and goal metrics) is as important as it is in simulation and closed-course testing. However, road testing depends on the maturity achieved in previous test formats, meaning that the interfaces and functionalities need to be established before road testing occurs. These tests can be gradual or part of an ODD expansion schedule in which the prototypes can be exposed to portions of the ODD up to the full extent. Similarly, in the development of an aircraft, the flight test campaign is restricted in its envelope (similar to ODD), according to the level of maturity of systems. Normal operation, selected safety culture, and degraded operational modes can be verified in this phase. From the program management perspective, having more than one prototype can accelerate road testing.

Testing Strategy As part of the testing strategy, the verification plan will define the types of tests to be performed, how their evolution and interdependencies will work in terms of ODD expansion, and resources needed. This is a gradual process that builds from the bottom up, from item- to product-level in an iterative, overlapping process that accounts for updates and corrections. The challenges of balancing these methods of testing are discussed in depth in the SAE EDGE™ Research Report "Unsettled Issues in Balancing Virtual, Closed-Course, and Public-Road Testing of Automated Driving Systems."

Scenarios Scenarios play an important role in the overall design of the autonomous vehicle and autonomous system [26, 27, 28] for specific reasons:

- They help on the definition and validation of the safety objectives.
- In aerospace, scenarios are used in simulations, testing, and labs for integration of systems and risk mitigation. They also provide the basis for failure conditions, as described in FHAs, and validation, which are used to identify the criticality of the integrity and availability of functions and to establish safety requirements implemented in system designs.
- They define expected behavior and performance of product and systems.
- They indicate how credit is taken from the driver, therefore helping to define what is expected from the autonomous system.
- They are an important aspect for testing by simulation, especially when it concerns the test of corner cases or those with high safety risks.

Some companies invest resources in developing their own automatic scenario-generation tools. Others purchase them from third parties.

Similarly, the aerospace industry, which is moving toward embedding autonomous systems into aircraft, will need to use the "scenario" concept in their development, considering that an aircraft might now be expected to operate autonomously. These autonomous operations may occur on the ground at airports and in flight. Even then, the core concept is the same and the scenario generation concept discussed here may apply to aerospace as well.

Recommendations

The understanding of how a human driver's characteristics influence the act of driving is an important first step in defining the design scope (requirements) an autonomous system has to comply with. The suggested regulatory text is an attempt to capture those requirements at a high level, which we will break down into executable engineering activities.

Since automotive (autonomous vehicles) and aerospace have several similarities, the suggested regulatory text, apart from the obvious differences (pilot or driver, flight envelope or ODD), should be applied on both fields' regulatory documents after review and agreement with regulatory bodies.

There are independent efforts from the aerospace and automotive fields to define a framework for the safety clearance of their specific products. However, due to the level of similarities in product functionality, operational characteristics, and safety objectives, it would be advantageous for the autonomous technology development in both fields to perform a joint discussion to shape a viable framework for the certification of products with embedded autonomous systems.

The industry, together with the traffic authorities and certification bodies, needs to agree and define on a minimum number of cases, or scenarios, to be tested and their proportional split between simulation, test tracks, and road testing. It is clear that accumulating driven miles is not solely sufficient to build technology credibility or prove safe behavior.

Safety Assessment Process

ARP4761 and ISO 26262 use similar approaches to define the criticality level. ISO 26262 uses "time exposure" - addition to effects on driver, passengers, and pedestrians - to help define the criticality level and safety objective. ARP4761 - said to be more qualitative - focuses mainly on effects on crew, passengers, and aircraft/system and establishes initial probability budget values to each failure condition. It is a qualitative analysis which utilizes quantitative probability to substantiate that safety-critical failure conditions have been appropriately mitigated.

The usage of exposure is part of the definition of the Automotive Safety Integrity Level (ASIL) classifications. However, equipment reliability requirements create different portability requirements than ARP4761 would. This is the difference between the two approaches which affects the design. ARP4761 is more "conservative" since it bases its demands for related work only on the effects of the failure conditions and associated design assurance level (DAL).

An aircraft product safety assessment is performed to meet the regulation requirements from the 14 CFR FAR 25.1309 (in this case Part 25 refers to transport aircraft - commercial planes). ARP4761 describes the guidelines for the safety assessment process of aircraft (product) and systems used as an acceptable means of showing compliance to the requirement. Similarly, ISO 26262 defines the safety process for automotive systems.

ARP4761 Approach

For ARP4761, the safety assessment starts at the product-level functionalities (or their hazard analysis). Table 1 lists, for clarity and comparison, aircraft functionalities and correspondent autonomous vehicles functionalities. Product-level functionalities are broken down into system-level functionalities. We will see in an example that more than one system can contribute to the performance of a product-level functionality.

A product-level FHA is performed and broken down into system-level FHA. For example, an aircraft ground deceleration is performed by the brakes, ground spoilers, and thrust reverser - each one part, respectively, of the brake system, flight controls, and engine. In a vehicle, similarly, we have the main brake system, engine, and manual braking lever (in older vehicles) contributing to vehicle braking function that can be used in conjunction to perform braking (Figure 14).

The FHA defined in ARP4761 analyzes the effects on cockpit crew, aircraft, and, passengers resulting from a function misbehaving or being lost (unavailable) and defines, based on those effects, the criticality of each case, or "failure condition," at different flight phases (i.e., taxiing, rolling takeoff, take off, climb, cruise, approach, and landing). Analyzing failure conditions at each phase of flight generates a list of scenarios from which the most critical are selected to define safety requirements (Figure 15).

It is worth noting the analogy with ISO 26262, where failure conditions relate to the safety cases, severity classification with safety goals, and flight phases to scenarios. A difference between the approaches is that the criticality definition method based on ARP4761 is derived from the effects of the failure condition, which will be translated into minor, major, hazardous, and catastrophic hazard levels. Those effects define the level of rigor of the processes to be adopted, the design protections, and probability targets,

TABLE 1. A side-by-side comparison of autonomous vehicle and aircraft functions is the starting point for the definition of the functional hazard assessment. It is worth noting that the DBWS aggregates the aircraft flight control, ground steering, and automatic flight control similar functionalities to provide direction and attitude control.

Aircraft functions (Product level)	Autonomous vehicle functions (Product level)
a. Flight control	a. Drive (directional) control
b. Ground steering	b. Drive (directional) control
c. Aircraft aspects of ATC	c. Autonomous vehicle aspects of GTC
d. Automatic flight control	d. Drive (directional) control
e. Cargo handling	e. Cargo handling
f. Engine control	f. Engine control
g. Collision avoidance	g. Collision avoidance
h. Ground deceleration	h. Autonomous vehicle deceleration
i. Environmental control	i. Environmental control
j. Passenger comfort	j. Passenger comfort
k. Communication	k. Communication
l. Guidance	l. Guidance
m. Navigation	m. Navigation
n. Passenger safety	n. Passenger safety

FIGURE 14. **Relation between product level FHA and systems level FHA (refer to ARP4761 Figure 2).**

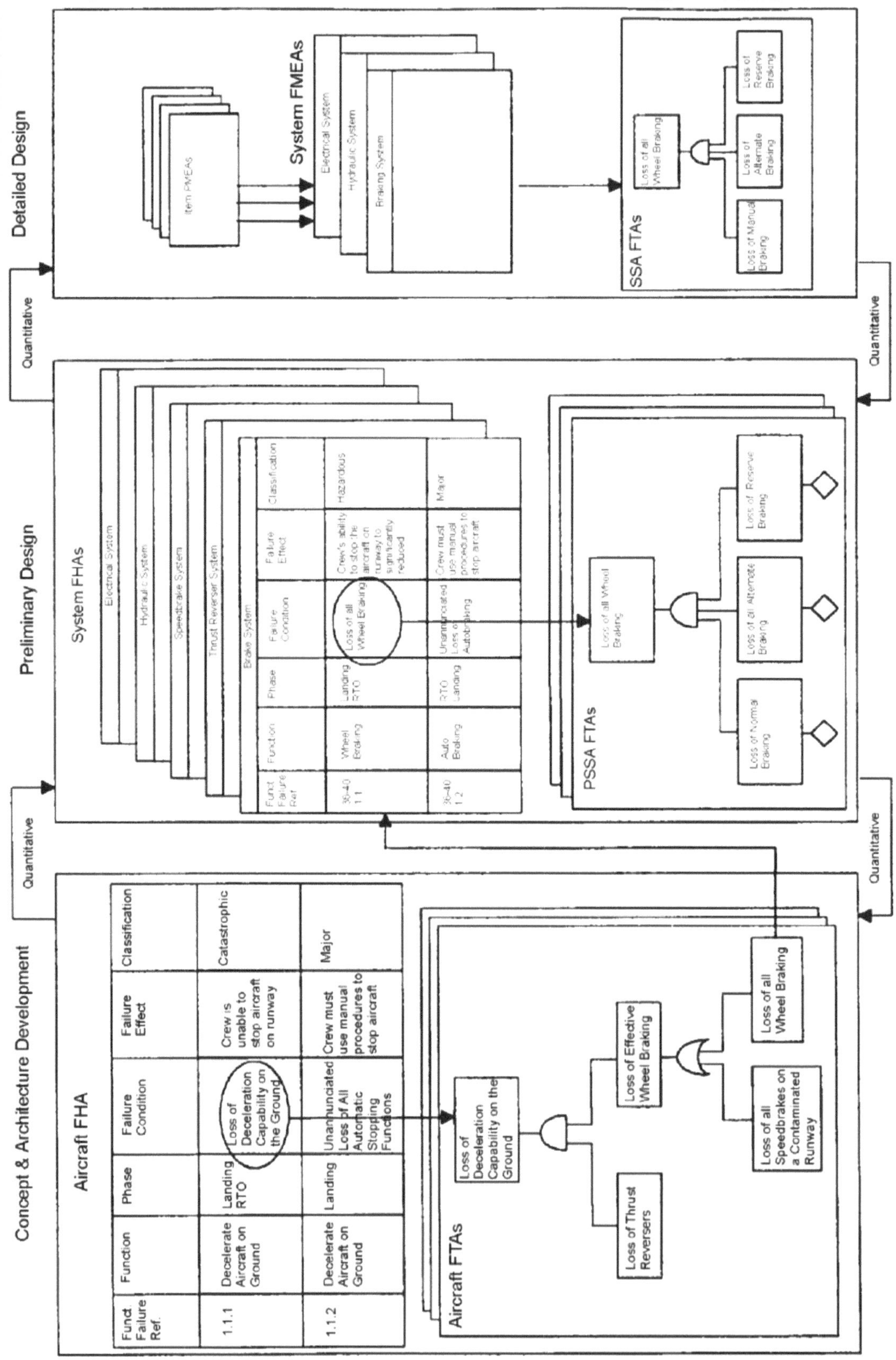

FIGURE 15. **Failure condition severity as related to probability objectives and assurance levels (refer to 25.1309-1A).**

Probability (Quantitative)	Per flight hour	1.0	1.0E-3	1.0E-5	1.0E-7	1.0E-9
Probability (Descriptive)	FAA	Probable		Improbable		Extremely Improbable
	JAA	Frequent	Reasonably Probable	Remote	Extremely Remote	Extremely Improbable
Failure Condition Severity Classification	FAA	Minor		Major	Severe Major	Catastrophic
	JAA	Minor		Major	Hazardous	Catastrophic
Failure Condition Effect	FAA & JAA	- slight reduction in safety margins - slight increase in crew workload - some inconvenience to occupants		- significant reduction in safety margins or functional capabilities - significant increase in crew workload or in conditions impairing crew efficiency - some discomfort to occupants	- large reduction in safety margins or functional capabilities - higher workload or physical distress such that the crew could not be relied upon to perform tasks accurately or completely - adverse effects upon occupants	- all failure conditions which prevent continued safe flight and landing
Development Assurance Level	ARP 4754	Level D		Level C	Level B	Level A

Note: A "No Safety Effect" Development Assurance Level E exists which may span any probability range.

which translate into reliability numbers for components and their failure modes. The fault tree analysis helps to check the reliability and required independence of the aggregated failure modes related to equipment contributing to a failure condition. The design assurance process makes sure that each failure condition is properly analyzed.

ISO 26262 approach

For ISO 26262, the safety goal is defined based on the effects on the driver, passengers of the vehicle causing the hazardous event, and other persons potentially at risk such as cyclists, pedestrians, or occupants of other vehicles. This is measured in terms of severity, exposure, and controllably. Then a probability number, ASIL, is defined (ISO 26262 - Part 3, Table 4).

where

- S1 to S3: Severity (health impact on the driver)
    - S1: Light and moderate injuries
    - S2: Severe and life-threatening injuries (survival probable)
    - S3: Life-threatening injuries (survival uncertain), fatal injuries
- E1 to E4: Exposure (of the driver to the risky situation)
    - E1: Very low probability
    - E2: Low probability
    - E3: Medium probability
    - E4: High probability
- C1 to C3: Controllability (by the driver)
    - C1: Simply controllable
    - C2: Normally controllable
    - C3: Difficult to control or uncontrollable
- QM: Quality Management
- A, B, C, D: ASIL A, B, C, D

Comparison of Approaches and Adjustments

The safety assessment of the autonomous vehicle and autonomous system (AI system and DBWS) must consider both approaches mentioned above. From the aerospace perspective, the assessment of controllability and severity is of utmost importance, but not the "exposure" factor, as it is believed to be misleading and causes lowered reliability requirements and less rigorous processes. Exposure refers to the fraction

FIGURE 16. **Breakdown of a total loss of brake functionality.**

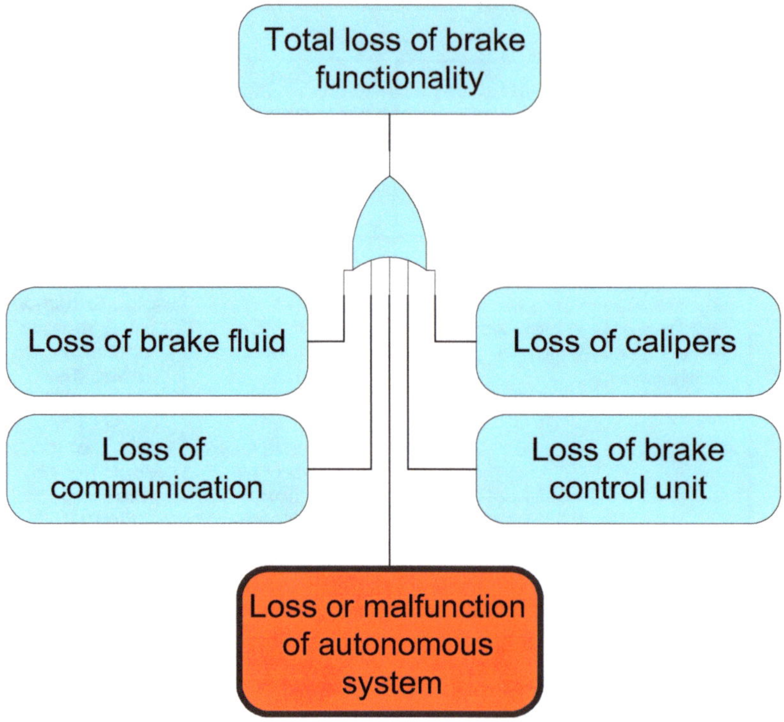

of time a failure might occur and affects the final probability of a safety case.

Without downplaying the importance of probability, a common phrase in the aerospace field is that "We do not design safety by numbers." However, looking back at the criticality definition, some adjustments in terms of the objects suffering from the effects of a malfunction need to be made to both approaches when applied to an autonomous vehicle and autonomous system.

It is important that we understand what impacts only the autonomous vehicle and autonomous system when deriving requirements from failure conditions or safety cases (i.e., "What safety requirements will impact autonomous vehicle and system?"). For instance, if total loss of brake functionality might be caused by the autonomous system as well as by any other component inside the brake system, then what is of interest are requirements related to the autonomous system only, including its interfaces, or the failure modes associated with it. Then, a coherent definition of requirements to the autonomous system can be derived. Figure 16 provides an example of the failure condition "total loss of brake functionality" displaying a fault tree with possible contributors.

Effects on cockpit crew (pilot and co-pilot) determine the level of workload and the capability of the crew to cope with an adverse situation. Considering the autonomous vehicle; whatever workload demand was needed for the driver (or pilot in an aircraft) in the standard vehicle now falls on the autonomous system. Effects on passengers can be expanded to include pedestrians and other objects. Effects on aircraft must be understood as effects on autonomous vehicles and other vehicles that might be affected.

Similarly, on the definition of the Severity and Controllability indexes (we are excluding Exposure), we must consider the autonomous system replacing the driver. The flight phases in the aerospace approach must be understood as the operational situations and operating modes, or scenarios, as dictated by ISO 26262 (part 3, clause 6 - 6.4.2).

The aerospace approach for severity classification defines the DAL for the function under analysis. The ISO 26262 approach is based on safety goals. There are two important aspects to understand:

1. Assess the autonomous system via both approaches and compare their outcomes in terms of work products to generate and process rigor to apply.
2. The DAL approach for autonomous systems has to be mapped into the ISO 26262 approach for automotive subsystems.

Therefore, there needs to be a translation from DAL to ASIL in place not based on probabilities. A suggested

TABLE 2. DAL to ASIL mapping.

Mapping	
DAL	**ASIL**
A	D
D	C
C	B
D	A
E	QM

mapping is given in Table 2, even though there are differences in their approaches and methodologies as seen above [29]. In cases when one approach is more demanding than the other, a decision needs to be made on which one is more valuable even if the workload increases.

ARP4754A and ISO 26262 indicate the work products to be generated according to the criticality assigned to failure conditions and safety cases, respectively.

A Practical Case What if the equipment supplied by a third party does not fulfill all of an autonomous system's safety requirements, or the DAL from a certain function does not correspond to the ASIL level supplied by the automotive system?

The solution depends on the limitations of the supplier equipment. In general, time and cost have to be compared against these options:

- Change software in supplier's control units for performance improvement, addition of functionality and/or addition of monitoring (e.g., extra sensor(s))
- Resolve the issue at the autonomous system level

The example in Figure 8—where there are sensors directly connected to the autonomous system—would require resolving the issue at the autonomous system level. But this solution also needs to undergo V&V for lost redundancy and integrity.

How Scenarios Help Safety Analysis

The definition of scenarios is of utmost importance in several steps of the analysis and development of the autonomous system. The discussion on how to approach the safety aspects and requirement definitions for the autonomous system suggests a merged approach to the definition of the hazards (safety objectives) for the failure conditions (safety cases), tailored to add value to the analysis. This approach looks at the safety classification from two perspectives which is beneficial for the definition of the autonomous system protective features. ARP4761 addresses electric/electronic and mechanical failures modes, while ISO 26262, based on IEC 61508, only addresses electric/electronic failures modes.

A coherent definition of safety objectives supports the verification via testing of the implemented requirements. From the safety requirements the following features are defined:

- Fail-safe devices: when failed do not incur in safety issues
- Independence: two different functions are independent in terms of data sources
- Segregation: two similar functions running on different platforms
- Dissimilarity: different hardware and software performing same tasks
- Redundancy: a secondary means to continue performing a function should the first system become inoperative
- Monitoring: data monitoring
- Voting: data selection or averaging

Each item from the list requires specific V&V methods.

This section describes how the aerospace approach to perform safety can be merged (combined) with the ISO 26262 approach. This new, more-complete framework helps to illustrate how functions are broken down from product-level to item-level and are used to identify safety systems for each level. This is an important process to derive safety requirements for further verification.

Example: Loss of deceleration in a scenario where there is a clear collision with a pedestrian is catastrophic per the aerospace interpretation. This can be caused by failure in different systems:

- Sensors, which did not detect an object properly
- AIS did not process the data correctly
- DBWS did not issue a brake command
- The Brake Control unit did not issue a brake command
- There is no brake fluid enough to provide correct levels of deceleration (leakage)

The list is just an example of some possible failure modes, and they reveal the variety of sources of failures contributing to the event. By assessing the failure modes, one is able to understand the need for the fail-safe features in the design defined by safety requirements. Processing data correctly either relates with the need for voting (at software level) and/or dissimilarity (of software and hardware). Not knowing that there is leakage in the brake system fluid implies the need for monitoring.

Consequently, a suggested strategy for verification at system and item levels is to test if those protections are in place by injecting data into the buses and checking how the voters would react or by purposely lowering the pressure in the brake system lines to check if the monitors would detect it.

The work products of this merged process captured in a safety plan are

- Autonomous vehicle FHA
- Systems FHA
- Safety requirements

Assessing Different Products

Is performance of the safety assessment different between products designed from the ground up and products where the autonomous system is being integrated into an existing platform? Is there any difference in testing for each case?

New Product Development

For the case of novel designs where vehicle, automotive systems, and autonomous systems are being designed from scratch, there is a need for a full safety assessment at all levels. In practice, automotive system solutions are available and then the process is to review each system safety documentation and understand what driver tasks will be automated. These will establish the safety requirements for the autonomous system.

For example, if the selected steering system is used in another vehicle, the existent safety documentation must be reviewed to assure it meets the safety requirements defined for that system with an understanding of safety cases and automated tasks. Is credit taken from a driver for reaction time or situational awareness? Those questions uncover the safety requirements for the autonomous systems (e.g., monitoring and processing times might impact safety).

In terms of verification, even if the autonomous systems are used in other vehicles, little to nothing is left for credit from previous applications of the automotive systems and the verification planning must take into consideration end-to-end testing as well as each individual system. Even though one might consider a similarity analysis, for automotive systems, it may or may not be accepted as MoC to a safety requirement. This will depend on agreements with the certification authority.

Integration of Autonomous System to Existing Platform For existing platforms with an integrated autonomous system, the safety cases need to be reviewed and automated driver functions must be understood.

The existing automotive systems might be limited in the information they make available to the autonomous systems, requiring the addition of other components to cope with the desired redundancy or integrity.

In terms of verification, focus must be on the integration of the autonomous system to the automotive platform and end-to-end testing. Performance testing is also required, but not to the same rigor as in the previous case. Tests performed on the original platform can carry over if they concern automotive systems integration and performance, though this is at the discretion of each manufacturer.

Validation and Verification Processes

In order to better discuss the object of this report, it is important to define V&V processes related to highly integrated, complex systems.

Validation Per ARP4754A, validation is the process that checks if a requirement is correct and complete, and the guideline defines how to proceed to check those two aspects. In short, it needs to know if the right aircraft (product)/system/function/item) is being built to meet the needs of customers, users, suppliers, maintainers, and certification authorities as well as aircraft (product), system, and item developers.

In the validation process, an initial matrix of requirements is passed through correctness and completeness checks, according to the validation rigor defined by the safety hazard criticality assigned to the requirement based on the failure condition. By passing those checks, a final validation matrix is produced with any deviation captured in a problem report for further analysis.

Validation methods include analysis, modeling/simulation, testing, similarity (service experience), and engineering reviews.

More than one validation method must be applied according to the safety criticality identified by the DAL. This variation of validation rigor based on safety-hazard criticality is carried down to the software and electronic hardware development processes, which support system functionality.

Verification The verification process checks if the implemented validated requirements meet their intent. Similarly, before using the final validation matrix as an input, follow the verification plan and check those implemented requirements based on the verification rigor assigned according to the requirement safety-hazard criticality and produce a final verification matrix. The deviations are captured in a verification summary, and problem reports are opened should that be the case.

Verification methods include inspection, review, analysis (of simulation results), testing, and service experience.

Like validation, more than one verification method is applied according to the safety criticality identified by the DAL. This variation of verification rigor, based on safety-hazard criticality, is carried down to the software and electronic hardware development processes, which support system functionality.

The Ethical Question and Moral Dilemmas

Articles and forums discuss the ethical and moral dilemmas that a self-driving car might face in performing its functions, such as in an unexpected critical situation involving the decision to sacrifice either its passengers or pedestrians. There are some aspects in the development of the AI system and DBWS which can be used to explain how those concerns can be addressed, how they are less likely to occur, and how they may be completely eliminated.

In the development plan, one of the sources of requirements is the list of scenarios we would expect the autonomous vehicle to be exposed to. These scenarios translate into safety margins embedded in the system hardware as fail-safe devices and software. These will provide layers of protection before something can go wrong. Should the development process guarantee the correct implementation of these features, then no question needs to be answered since these types of problems are resolved by design. In other words, should an autonomous vehicle be exposed to a situation where it needs to decide between occupants and pedestrians because layers of protection have previously failed, the autonomous vehicle is potentially in a "total loss of control" condition where decision making is irrelevant and a catastrophic event is the likely outcome.

Autonomous vehicles do not make decisions like that: a safe trajectory is free of collisions.

Having a robust development process in place, based on assessment on FHA of all functional conditions (or safety cases per the ISO 26262) at all levels (product, systems and items) where everything contributing to a "total loss of control" of the driving functionality or an "unintended command" is looked at, is what will define safety requirements where redundancies, monitoring, integrity, reliability, fail-safe features, etc., are defined. The whole point of the requirement-based design development process, with the support of the validation of such requirements and their verification, is to make sure no unsafe condition with injuries of fatal consequences would ever happen. This is the important point of this report, where the development process is explained. Safety is not only attained by physical characteristics embedded into the hardware, or interlocks in software, but by the intense and extensive process that defines and checks all requirements.

Recommendations

Adopt ARP4761 (together with the ARP4754A) for functional DAL breakdown to define the hazard levels given that effects are adjusted to the self-driving context for the autonomous vehicle and system. It provides a conservative methodology for the definition of safety requirements considering it has been successfully applied in highly integrated and complex systems, such as fly-by-wire ones in an aircraft. In that sense the ARP4761 is better equipped to deal with this challenge [30].

Support the safety assessment with safety cases, using the ISO 26262 approach, to identify the safety critical ones.

Derive the safety requirements for the autonomous vehicle and system, considering what driving tasks are being replaced by a computational system.

Perform other safety-related processes to the autonomous system such as common cause analysis (encompassing Zonal Hazard Analysis, Particular Risks Analysis, and Common Mode Analysis).

From the safety requirements, derive the verification strategy for the autonomous vehicle and system, which will unfold from understanding the effects of losing any functionality allocated to the autonomous system or losing this system as a whole. This also includes deriving the effects of system malfunctions that cause unwanted commands and erroneous data. These tests will probably be developed under the degraded operational mode, to verify either that proper redundancy and integrity are built-in, or that all possible fail-safe features protect the autonomous vehicle from critical hazards.

Control Architecture - Hardware

The hardware control architecture concept and design are a result of several types of requirements, safety being the most important one. Not understanding what we are designing against or selecting an ill-fit methodology affects the final product as the shortcomings might not be detected during the V&V of implemented requirements.

The hardware control architecture for highly-integrated and complex systems in aerospace is based on the basic concept of channels comprising several dissimilar command and monitor lanes to provide proper integrity and redundancy. This type of architecture is proven to work, since it is flying in several different airplanes and has been suggested to be the core of the autonomous system [26].

The control architecture is assessed as a whole and has the responsibility to fulfill the safety requirements. The responsibility cannot be allocated to one component alone, as a multicore processor with several applications in **automotive** systems, or to any of its embedded functionalities to check correctness of code, lockstep functionalities, etc. However, due to its processing capabilities, it can be used in a control architecture, given it is well configured (constrained in some features as lockstep) and verified for such applications, which is important due to lack of determining characteristics [31, 32, 33].

Consequences of not having a proper control architecture are

- Incapability of meeting safety requirements
- Incapability of meeting operational requirements or presenting a competitive operational behavior

The V&V processes cannot help in the certification of a control architecture if their options and their consequences are not well understood.

Products and Businesses

Tesla's Autopilot system, and Waymo's product are examples of autonomous solutions for different market segments. According to the operational profile of the Tesla Autopilot system, the system needs to be of high integrity to avoid making a mistake while driving, redundant enough should one computer fail, and capable of alerting the driver to give back control in case of a failure.

On a fleet of robot-taxis, integrity is an important characteristic for the same reason as the Autopilot, and so is redundancy. If its operational profile defines a non-interruptible service, it would be annoying to interrupt a ride because either the computer, AI computer, or DBWC is inoperative. Redundancy of sensors is important to guarantee area coverage in the event of loss of one or more types of sensor (considering LiDAR, radar, and camera). Safety and operational requirements are important to define the control architecture for AI system and DBWS (Figure 17).

FIGURE 17. **High-level sources of requirements define the lowest level of requirements at the item level.**

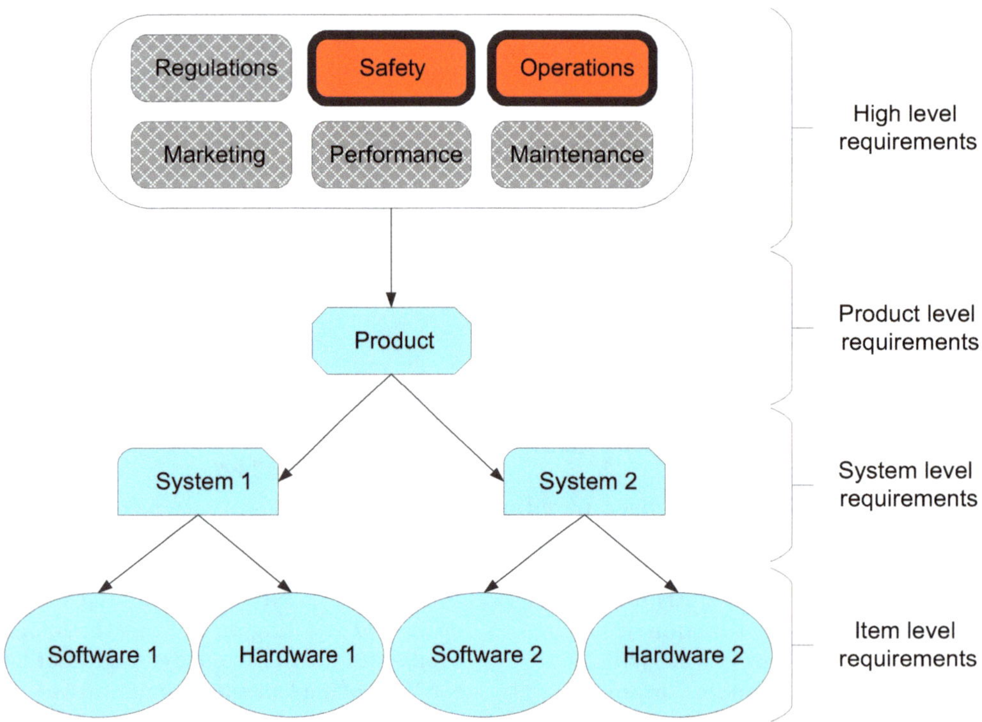

Verification of Control Architecture

At the system level, the procedure to test redundancy is as follows:

1. Inject failures and verify if the proper data voting performs as requested.
2. Interrupt power to one computer channel (e.g., that would make a redundant computer fail-silent) and check that the adjacent computer resumes control without creating any transience in data and commands. In Figure 18, for example, it would mean disabling one channel and checking that the adjacent channel resumes control or vice-versa.
3. Shut down subsets of sensors and check performance and operation under degraded vision.

Again, these tests must be performed in the context of scenarios. At the item level, the following tests can be performed, considering the example in Figure 18:

1. Erroneous data in the data bus being consumed by all computers to check data voting
2. Failure injection in the cross-lane data links
3. Failure injection in the cross-channel data link

The control architecture in Figure 18 is a standard considering one dissimilar processor per lane. Dissimilarity of processors protects against generic random failures from manufacturing. There are alternatives for the number of dissimilar processors per lane (i.e., the main concept of having a processor performing as a "safe gate"), usually the monitor lane processor with higher DAL, A, and a command processor executing the actions, usually a lower DAL, C (based on ARP4754A logic). The DAL functional breakdown is explained in the ARP4754A and previous versions.

Recommendations

Implement a high-integrity and redundant control architecture in both AI system and DBWS with similar concept, command, and monitor features, as applied in aerospace safety critical systems. It is a traditional concept and very reliable solution used in aerospace that has been proven over millions of flights across countless aeronautical miles. It gives

FIGURE 18. Generic Command and Monitor Control Architecture: Two channels with dual lanes.

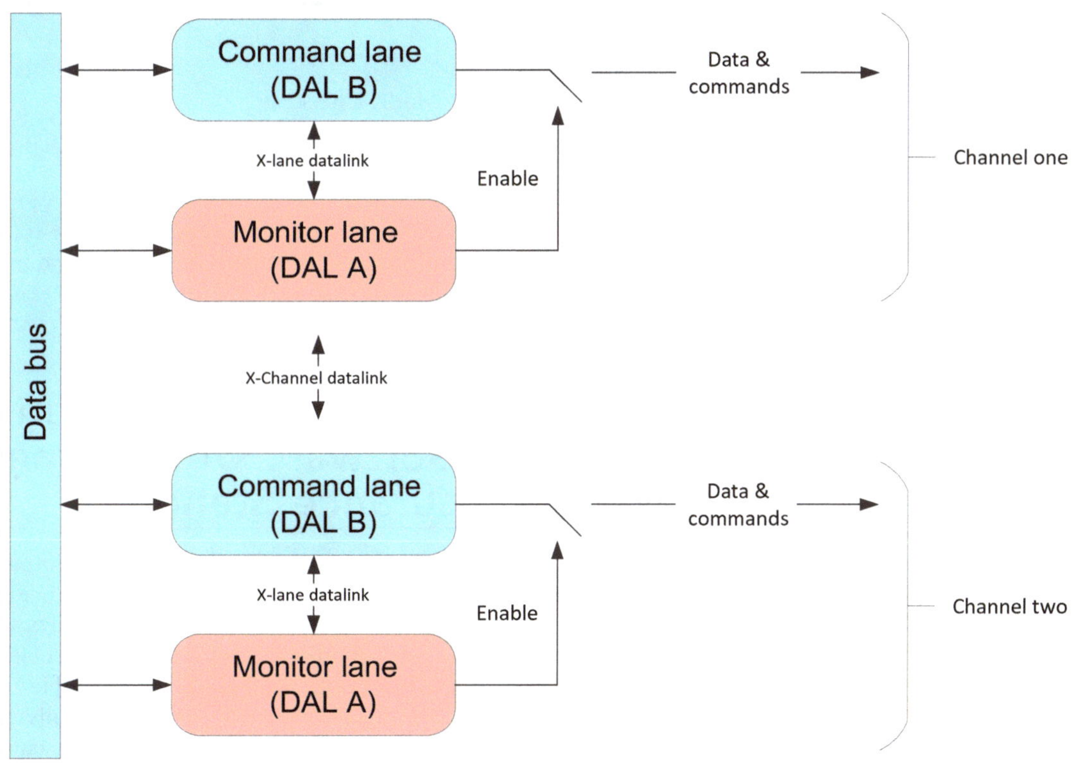

"legitimacy" to this new autonomous technology field. This architecture guarantees the integrity of data processes in the AI computer and correctness of commands issued by the DBWC, considering an AI system only processes sensor data and DBWC executes critical commands.

Consider the experience and existing FAA guidelines concerning the applicability of multicore processor in the development of autonomous systems control architecture.

Forbid the usage of a lockstep feature in a safety-critical application, such as autonomous vehicle, since it implies processing the same code and comparing line by line. This goes against one of the important and desired features of autonomous software, which is to be dissimilar as will be seen in the further section.

Software V&V Challenges

There is a requirement flow from the product level to item level (software is an item in ISO 26262). Therefore, the decomposition of higher-level requirements into item-level ones is important. A wrong requirement, not necessarily incorrect, can pass through the funnel of the process, therefore indicating the importance of the sections above on how to approach safety and define correct requirements and the definition of regulation requirements.

In an autonomous vehicle, simply put, AI is used in the data processing phase to generate an interpretation of a 3D world, the definition of trajectories, and the overall driving functionalities. AI software has been applied in different fields including in safety-critical applications, for example, in the Remote Agent Experiment (RAX) from NASA [34]. However, its verification, due to some of its characteristics further explained, still presents challenges due to some of its intrinsic qualities [26, 34].

The inability to fully verify AI software prevents us from confirming its compliance to safety requirements, and this factor affects the deployment of products and services in either their entirety or parts, in the form of restrictions on the ODD. However, there are promising suggestions to perform V&V on AI software.

To understand the challenges imposed and possible solution to the verification problem, it is important to know basic AI software characteristics, trends to verify it, and the context of software certification process. It is also important to understand how the RTCA-DO-178 framework (and/or others)—used to certify aerospace complex software functionalities—can eventually help ISO 26262 software verification and work product generation.

This section will not discuss in depth the technicalities of AI software V&V, but intends to inform about the current landscape.

Nondeterministic Nature of AI Algorithms

The functionalities applied in an autonomous vehicle/system are nondeterministic in nature. They have probabilities associated with them (e.g., selecting the correct path, identifying correctly objects, and determining if those will be in a collision path with the vehicle) [26, 34].

Testing of such systems becomes difficult for two reasons [26]:

1. The difficulty of exercising a specific edge-case situation
2. The difficulty of evaluating whether test results are correct or not, because there is no unique correct system behavior for a given test case

It is more important to verify the statistical characteristic of the behavior rather than its correctness. Also, machine learning using inductive learning, for better parameter-tuning efficiency, poses challenges to verification in terms of definition of requirements to test (they fall out of the traditional "V" format), data used for testing (training data), and the extra data amount required for re-testing the result "legibility" problem imposed to humans and the "black swan [26]."

Nevertheless, AI can deliver much more functionality at a reduced cost. It poses challenges to be verified by conventional methods; however, there are suggested readily available methods that enable the V&V of complex AI software [34]:

- Run-Time Monitoring
- Static Analysis
- Model Checking
- Theorem Proving

Also, there are important features of an AI software application to be understood, such as its complexity, declarative knowledge or model-based, and the approaches to verify each one, making for available a range of alternatives [33].

Cross-Domain Comparison of Software Development Assurance Standards

In the attempt to compare and define a tailored solution that supports the verification and safety clearance of an autonomous systems, it is important to understand how the different software development standards prescribe objectives, activities, and final work products for comparable DALs.

Does the DAL mean the same for the different standards? This is the first question to answer before comparing what activities need to be performed and what work products need to be generated.

The product- and system-levels define DALs for software item definition.

In Ledinot [35], this effort is captured and discussed in detail. Several standards are analyzed (DO-178/ED-12, ISO 26262, IEC 61508-3, CENELEC EN 50128, ECSS-Q-ST-80C, etc.) and are cross compared in terms of their

- Quality assurance objectives
- Process activities
- Process work products
- Development means
- Independence of some verification activities
- Content of the software product

The outcome is that there are significative foundational differences among the standards which do not help in the attempt to tailor a solution for the autonomous system software process definition in support of its design.

Recommendations

Koopman [26] and Pecheur [34] describe the characteristics of AI software and the challenges for their verification and available alternatives. The AI software for autonomous systems could be assessed through these criteria to define what better technique there is to perform verification.

In terms of the software design process, it is greatly affected by how each standard is defined as seen in Ledinot [35]; therefore either a robust standard with years of experience (aeronautics, nuclear, or space) is used in supporting the software processes or a more tortuous way is defined and a tailored solution is agreed upon, which implies great time and cost. A midpoint approach might be to consider a traditional standard, such aerospace, and work out some points that need extra attention.

The suggested solution in Koopman [26] to address the nondeterminism of the software's high-critical functionalities is via synthetic data, or simulation to handle high volume of data needed, and a "dissimilar" software architecture, based originally on ASIL decomposition. We recommend the DAL decomposition and not the ASIL. Software dissimilarity, as per aerospace practice, implies—in different language/compilers—implementing the same set of requirements by different teams. For AI code this can be expanded to different AI algorithm approaches running in the command and monitor lanes of the AI computer and DWBC.

Conclusion

Autonomous systems embedded with AI algorithms are a new technology currently lacking design-regulation requirements, therefore making it difficult to understand what is required and how to efficiently apply resources to infrastructure and engineering activities. This leaves it up to each player to define individual requirements, which leads to divergent solutions for designing safety features in autonomous vehicles. Different interpretations of safety needs might generate diverse hardware and software solutions, creating a risk of not covering actual safety concerns, and potentially putting the credibility of the autonomous technology and test campaigns at risk.

Therefore, for all reasons stated above, it is of utmost importance to define a framework to support the development and certification of autonomous products and systems. This report suggests a regulatory text capturing the core needs for autonomous products and systems, which simply looks at what characteristics of the human driver the autonomous technology must replicate with equal or better performance and how it drives the definition of verification plans and tests.

The definition of a safety process supported by aerospace guidelines opens space for a detailed analysis of the implications of having an autonomous system in control of a vehicle, providing a more conservative approach in the definition of safety requirements which can also be verified according to the integrity and redundancy solutions adopted. The suggested control architecture solution is aerospace-based with millions of aeronautical miles flown, reinforcing the safety protections needed for a safety-critical system such as the autonomous one is.

On software development of AI algorithms, the challenge is in its nondeterministic nature, making verification difficult. However, alternatives for verification strategies have been listed. This unsettled domain needs focused attention regarding the definition of proper design assurance and consequent framework. This deserves, in comparison to the other domains, a more focused attention.

All the items discussed influence the entire verification and test strategy. One will not progress or be successful without the proper outline of the main pillars (requirements definition).

The whole approach of using the aerospace experience, processes, frameworks, etc. can give the credibility and legitimacy the autonomous technology needs for public acceptance. Once the processes are defined and directions are given, the ethical question does not need to be posed as the concerns will be properly resolved in the design process.

SAE EDGE™ Research Reports

SAE EDGE™ Research Reports, like the present report on autonomous system V&V processes, are intended to push further out into still unsettled areas of technology of interest to the mobility industry. SAE launches these reports before attempting to form a joint working group, let alone a cooperative research project or a Standards Committee.

SAE EDGE™ reports are intended to be quick, concise overviews of major unsettled areas where vital new technologies are emerging. By definition, an unsettled area is characterized more by confusion and controversy than established order. Early practitioners must confront an absence of agreement. Their challenge is often not to seize the high ground but to find common ground. These scouting reports from the frontiers of investigation are intended merely to begin the process of sorting through critical issues, contributing to a better understanding of key problems, and providing helpful suggestions about possible next steps and avenues of investigation.

SAE EDGE™ Research Reports, therefore, are fundamentally distinct from the more formal working groups approach and far removed from the more mature cooperative research program and standard's development process.

Next Steps for Autonomous System V&V Processes

This publication should be considered only as a first step toward clarifying the issues around autonomous system V&V processes. The intention behind this and other SAE EDGE™ Research Reports is to start a dialogue among interested parties on important industry-wide topics that require further attention. The expectation is that these explorations of unsettled areas of technology will lead to the formation of working groups and, ultimately, committees that can address and resolve the issues they raise, producing a framework for developing a common vocabulary of definitions, best practices, protocols, and standards needed to support continued progress toward safer and more innovative automotive products.

The expert discussions that gave rise to this publication demonstrated a great willingness on the part of the industry to define the terminology, procedures, and eventually the standards needed to enable autonomous system V&V to move ahead as quickly and efficiently as possible.

This SAE EDGE™ Research Report on Unsettled Domains Concerning Autonomous System Validation and Verification Processes identifies the following key topics for further pursuit, both through continued informal discussions among industry practitioners and through more formal working groups:

- Invite the industry and collaborators for a broad discussion of the unsettled domains described here.
- Narrow, instead of widen, the discussions toward adopting proven frameworks and processes from aerospace and space in addition to the ISO 26262.
- Adopt a more conservative approach in the design of autonomous vehicles and systems not based only on the ISO 26262 standard.

Recommendations

The overall recommendations of this SAE EDGE™ Research Report can be summarized as follows:

1. Understanding of the awareness, decision-making, and action processes a human driver goes through helps in defining the requirements for autonomous systems in terms of performance, behavior, and monitoring. It also gives the minimum metrics for comparison of fail-success criteria.
2. Simulation must be discussed to aid the verification of safety critical cases and to prove that an autonomous vehicle and system is safer than a human driver. Scenarios can be built from NHTSA's accidents/incidents database, given adjusted vehicle characteristics, to verify that correct levels of safety, corresponding features, and expected behavior meets requirements.
3. Automotive and aerospace products and systems have several similarities in their functions and design processes which make it easy to translate concepts between fields. Even an airline company has similarities with the autonomous mobility services being designed. This is a good start for tailoring frameworks from one field to the other.
4. The agreement over the scope and minimum amount of testing using labs, simulation, test tracks, and road tests helps in the planning and resources investment in a development program.
5. The adoption of the aerospace framework for safety assessment brings value due to the approach used, effects rather probabilities only, covering any potential gap that ISO 26262 alone is not capable to answer. Since understanding safety cases and scenarios is important for the definition of what touches the autonomous product and system design is important, a joint tailored approach is suggested.
6. A good understanding of safety aspects of the autonomous vehicle and system helps in the definition of the features they need to have and the verification plan (testing).
7. A standard high integrity and redundant control architecture, as the one suggested in the text with dissimilar processors and redundant channels, meets the highest DALs defined for the AIS and DBWS, and establishes the same level of legitimacy and public perception found in aerospace and space products.
8. Multicore processors are making their entrance in the aerospace industry, they are more powerful in terms of performance than single-core processors; however, their nondeterminism raises concerns which are being researched by the aerospace community. There are guidelines and reports for their testing and usage, which can help a control-architecture certification and verification campaign.

9. AI software and its nondeterministic nature poses a challenge for certification of autonomous products, a correct definition of what a correct way of defining a design assurance-related process is still unanswered, even though there are some testing techniques already available that can help the verification process. It is important also to consider dissimilarity of code, avoidance in the usage of lockstep functionality, since in the case of an incorrect requirement being implemented in the code, the lockstep renders itself useless to help in that case. From the aerospace perspective, software dissimilarity (different compilers, algorithms approaches, and implementing teams) is nonnegotiable to guarantee integrity.

Abbreviations/Definitions

3D - Three-dimensional

AI - Artificial Intelligence

ANPRM - Advance Notice of Proposed Rulemaking. A letter NHTSA has released to the industry asking for contributions in the definition of a framework to support the achievement of safety clearance for autonomous vehicles.

ARP - Aerospace Recommended Practice

ASIL - Automotive Safety Integrity Level

DAL - Design Assurance Level

DBWC - Drive-By-Wire Computer. The processing unit, or computer, of the DBWS.

DBWS - Drive-By-Wire System. Processors, interfaces, sensors that translates given trajectories and field information into commands to automotive systems as brake, steering and acceleration. It also performs high level monitoring of autonomous vehicle modes and states.

DOT - Department of Transportation

FAA - Federal Aviation Association. The governmental body responsible for the rulemaking in the aerospace field

FAR - Federal Aviation Regulation

FHA - Functional Hazard Analysis

HARA - Hazard Analysis and Risk Assessment

LiDAR - Light Detection And Ranging

MoC - Means of Compliance. How a design item is going to be shown to satisfy a regulatory requirement. In general, the example of options is one or more of the following: tests, analysis, inspection, similarity with previous implementation.

NHTSA - National Highway Traffic Safety Administration

ODD - Operational Design Domain. A specified scenario in which a self-driving vehicle can safely operate on its own, e.g., highway driving.

RAX - Remote Agent Experiment

RTCA - Radio Technical Commission for Aeronautics

V&V - Verification and Validation

Acknowledgments

Recognition should go first to all the participants, many of whom also provided feedback on the draft version of this publication. Without their input and initiative, this SAE EDGE™ Research Report would not have been possible.

Alexandre Gesualdi, ACG Consulting BR
Paulo Allemand Donato, retired from Embraer
Robert Voros, Textron Aviation

The authors of this document together with the SAE Team responsible for its creation join in expressing our deepest appreciation to all the individuals mentioned above.

References

1. Zoox, "Zoox Online Page," www.zoox.com, 2018. https://zoox.com/.
2. Christofer, H., "What Can Self-Driving Cars Learn From Aviation? - The Drive," thedrive, 2017. https://www.thedrive.com/tech/13903/what-can-self-driving-cars-learn-from-aviation.
3. Schoenberger, R., "Aviation Industry Influences Autonomous Car Design - Today's Motor Vehicles," 2018. https://www.todaysmotorvehicles.com/article/autonomous-aviation-zenuity-adient-071118/.
4. Society of Automotive Engineers, "ARP4754A: Guidelines for Development of Civil Aircraft and Systems," Aerospace Recommended Practice, 2010.
5. RTCA, "RTCA-DO-178C - Software Considerations in Airborne Systems and Equipment Certification," 2012.
6. RTCA, "RTCA DO-254/EUROCAE ED-80, Design Assurance Guidance for Airborne Electronic Hardware," 2000.
7. Razdan, R., "Unsettled Technology Areas in Autonomous Vehicle Test and Validation," SAE EDGE Research Report, 2019.
8. Waymo, "Waymo Online Site," 2019. https://waymo.com/.
9. Cruise, "Cruise Automation Online Site," 2019. https://www.getcruise.com/.
10. Nuro, "Nuro AI Online Site," 2019. https://nuro.ai/.
11. Kodiak, "Kodiak AI Online Site," Kodiak Web Page, 2019. https://kodiak.ai/.

12. Starski, “Starsky Robotics Online Site,” Starski Robotics Web PAge, 2019. https://www.starsky.io/.
13. Uber, “Uber Elevate Online Site,” 2019.
14. Bell, “Bell Nexus Online Site,” 2019. https://www.bellflight.com/experience/on-demand-mobility.
15. Carl, E.B., “NASA Autonomous Operations Workshop,” 2019. https://www.faa.gov/news/speeches/news_story.cfm?newsId=23674.
16. Daniel, K.E., “Uber Elevate Urban Air Mobility Summit 2019,” FAA, 2019. https://www.faa.gov/news/speeches/news_story.cfm?newsId=23794.
17. SAE, “Surface Vehicle Recommended Practice - J3016,” 2018.
18. SAE, “ARP4761 - Guidelines and Methods for Conducting the Safety Assessment Process on Civil Airborne Systems and Equipment,” 1996.
19. ISO, “International Standard ISO 26262-3,” *ISO Standards*. Vol. 3. Geneva, 2018.
20. Tanenblatt, E. and Schneider, C., “Biggest Roadblock to Autonomous Vehicles Isn’t Technology,” 2019. https://www.wardsauto.com/industry-voices/biggest-roadblock-autonomous-vehicles-isn-t-technology.
21. Beiker, S., “Unsettled Topics Concerning Sensors for Automated Road Vehicles,” 2018.
22. NHTSA/DOT, “Pilot Program for Collaborative Research on Motor Vehicles with High or Full Driving Automation,” 2018.
23. Dentons, “Autonomous Vehicles: US Legal and Regulatory Landscape,” August 2019.
24. Araya, D., “The Big Challenges In Regulating Self-Driving Cars,” Forbes, 2019. https://www.forbes.com/sites/danielaraya/2019/01/29/the-challenges-with-regulating-self-driving-cars/#69f8dccbb260.
25. FAA, “Advisory Circular 25.1309-1A,” *Area* 1309, no. c 1988. https://doi.org/AFS-800 AC 91-97.
26. Koopman, P. and Wagner, M., “Challenges in Autonomous Vehicle Testing and Validation,” *SAE Int. J. Trans. Safety* 4(1):15-24, 2016, https://doi.org/10.4271/2016-01-0128.
27. Thorn, E., Kimmel, S., and Chaka, M., “A Framework for Automated Driving System Testable Cases and Scenarios,” United States, Department of Transportation/National Highway Traffic Safety, Technical Report, 2018.
28. Menzel, T., Bagschik, G., and Maurer, M., “Scenarios for Development, Test and Validation of Automated Vehicles,” in *2018 IEEE Intelligent Vehicles Symposium*, 2018, 7, arXiv:1801.08598.
29. Machrouh, J., Blanquart, J.P., Baufreton, P., Boulanger, J.P., Delseny, H. et al., “Cross Domain Comparison of System Assurance,” in *ERTS 2012*, Toulouse, 2012, 1–3 2012, 10.
30. Baufreton, P., Blanquart, J.P., Boulanger, J.L., Delseny, H., Derrien, J.C. et al., “Multi-Domain Comparison of Safety Standards,” in *ERTS2 2010, Embedded Real Time Software & Systems*, Toulouse, France, May 2010, ffhal-02264379f.
31. DOT/FAA, “Assurance of Multicore Processors in Airborne Systems - DOT/FAA/TC-16/51,” 2017.
32. FAA, “Multi-Core Processors - Position Paper CAST-32A,” 2016.
33. Parkinson, P., “Towards Avionics Safety Certification on Multi-Core Processor Architectures,” https://techtime.news, 2016. https://techtime.news/2016/10/31/processor/.
34. Menzies, T. and Pecheur, C., “Verification and Validation and Artificial Intelligent,” *Advances in Computers* 65:153-201, 2005.
35. Ledinot, E., Astruc, J.M., Blanquart, J.P., Baufreton, P., and Boulanger, J.L., “A Cross-Domain Comparison of Software Design Assurance Standards,” n.d.
36. “Uber Elevate: Uber Air,” Uber RideShare Site, 2019. https://www.uber.com/us/en/elevate/uberair/, accessed December 18, 2019.
37. Zoox, “Homepage,” 2019. https://zoox.com/, accessed December 18, 2019.

Contact Information

EDGEresearch@sae.org.

The authors of this document together with the SAE Team responsible for its creation join in expressing our deepest appreciation to all of the individuals who contributed.

www.ingramcontent.com/pod-product-compliance
Lightning Source LLC
LaVergne TN
LVHW070937210926
848482LV00003B/265
* 9 7 8 1 4 6 8 6 0 1 2 7 5 *